baby CAMELS

KIM THOMPSON

CREATIVE EDUCATION • CREATIVE PAPERBACKS

CONT

ENTS

I Am a Calf — 4

I Can Run! — 6

Desert Protection — 8

Helpful Hump — 10

Speak and Listen — 12

Camel Words — 14

Reading Corner — 15

Index — 16

I AM A CALF.

I am a baby camel.

I was born with my eyes open. I could run when I was just a few hours old.

My mom cares for me. She feeds me her milk.
We live in a group of camels called a caravan.

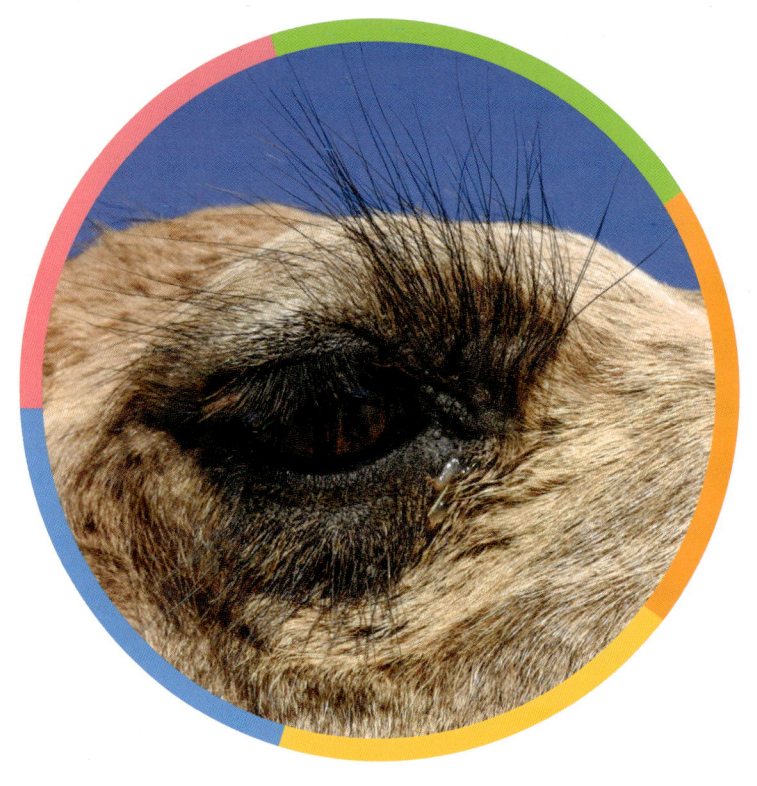

I live in the desert. Long eyelashes protect my eyes from the sun.

I can close my nostrils to keep out blowing sand.

My hump grows. It stores fat.

My hump can help me live for many days without food or water.

I will get big! I will stand as tall as a doorway.

SPEAK AND LISTEN

HAHHH!

Can you speak like a calf?

Baby camels cry and grunt.

Listen to these sounds:

https://www.youtube.com/watch?v=QPWSgQrfIIk

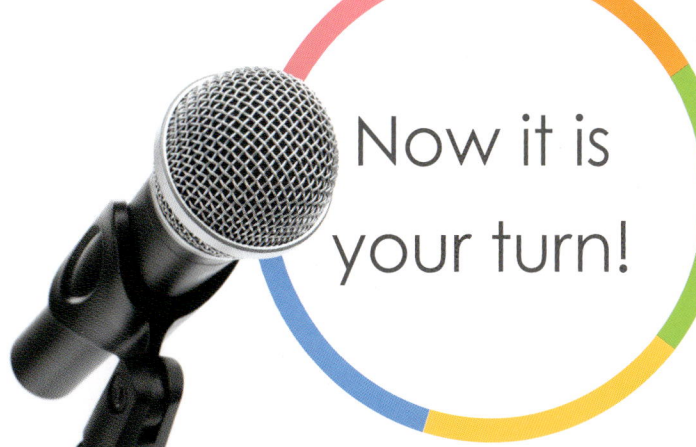

Now it is your turn!

CAMEL WORDS

desert: a dry area where hardly any plants grow because there is not much rain

hoof: a hard, nail-like covering on the foot of a camel, horse, or other animal

hump: the lump on a camel's back which stores fat that can be used as food

nostrils: openings in the nose that animals use to breathe

READING CORNER

Carryl, Charles. *The Camel's Lament.* London: HarperCollins, 2021.

Ganeri, Anita. *I Wonder Why Camels Have Humps: And Other Questions About Animals.* New York: Kingfisher, 2023.

Riggs, Kate. *Camels (Amazing Animals).* Mankato, Minn.: Creative Paperbacks, 2022.

INDEX

desert 8

eyes 4, 8

group 7

hump 5, 10, 11

lips 4

mom 7

run 6

sand 9

size 11

water 11

PUBLISHED BY CREATIVE EDUCATION AND CREATIVE PAPERBACKS
P.O. Box 227, Mankato, Minnesota 56002
Creative Education and Creative Paperbacks are imprints of The Creative Company
www.thecreativecompany.us

COPYRIGHT © 2026 CREATIVE EDUCATION, CREATIVE PAPERBACKS
International copyright reserved in all countries. No part of this book may be reproduced in any form without written permission from the publisher.

LIBRARY OF CONGRESS CATALOGING-IN-PUBLICATION DATA
Names: Thompson, Kim, 1970- author
Title: Baby camels / Kim Thompson.
Description: Mankato, Minnesota : Creative Education and Creative Paperbacks, [2026] | Series: Starting out | Includes bibliographical references and index. | Audience term: juvenile | Audience: Ages 4-7 Creative Education and Creative Paperbacks | Audience: Grades K-1 Creative Education and Creative Paperbacks | Summary: "Introduce beginning readers to the world of baby camels with this life science starter. Includes photos, a labeled animal diagram, "Make a Noise" section, glossary, and further resources"-- Provided by publisher.
Identifiers: LCCN 2024043237 (print) | LCCN 2024043238 (ebook) | ISBN 9781640264182 library binding | ISBN 9781628329513 paperback | ISBN 9781640005822 ebook
Subjects: LCSH: Camels--Infancy--Juvenile literature
Classification: LCC QL737.U54 T46 2026 (print) | LCC QL737.U54 (ebook) | DDC 599.63/62--dc23/eng/20250107
LC record available at https://lccn.loc.gov/2024043237
LC ebook record available at https://lccn.loc.gov/2024043238

DESIGN AND PRODUCTION
Design by Rhea Magaro
Production by Beeline Media and Design, Inc.
Art direction by Tom Morgan

PHOTOGRAPHS by Alamy Stock Photo/Joerg Drescher, 2-3; Shutterstock/abubakar0023, 11, Andrey Bocharov, 10-11, Bohbeh, 13, Homo Cosmicos, 12, J.NATAYO, 4, KYRA DUSHI, 9, LeStudio, cover, Lotta Axing, 6-7, malangusha, 7, photomaster, 5, SERSOLL, 14, Stefano Ember, 8

Printed in India